The Law of
Universal Mendacity

- and Don't Be Conned

By

Bo De Yang

波 德 杨

First published by AuthorHouse 04/08/04

ISBN: 1-4184-0572-8 (e-book)
ISBN: 1-4184-0573-6 (Paperback)

Library of Congress Control Number: 2003098970

Printed in the United States of America
Bloomington, IN

This book is printed on acid free paper.

CONTENTS

PREFACE

The law that keeps our feet on the ground is the same law of gravity that holds the moon high above the atmosphere. We don't avert to it to feel it. There are laws that govern the way we deal with information, though we may not notice them. We select information for relevance. Facts are rarely simple. In an age when lying is policy, those who indulge in dupery become sacrificial victims: most of all the naively loyal young men, of whatever persuasion, who put themselves in harm's way when their comfortable elders order it. When leaders lie they lose, eventually, their power to lead, because electorates do not like being deceived and the truth often comes out in time.

This scripted antidote is for responsible and democratic behavior. It is against being deceived - a trick vital to the military, truth being a counter revelation to the art of surprise and therefore the first casualty before enemies are engaged. Cheaters prosper because they take the advantage and are safe because the power of the honest is disadvantaged. This book is not overtly about truth and honesty, qualities much prized by some, qualities often instilled by education, and about which much has been written that can be taken for background. The book is tolerant of realistic failures of rationality in humankind. You are recommended constantly to move forward - no less when expectations are disappointed or surprised.

There are times when information has the advantage and when honesty can be enforced, but how do you ensure enforcement when you don't have an independent source for knowledge? Normally, you cannot make grown people honest. This book is for the individual: how you draw information from what other people say or write, whether honest or lying or unpredictable. A perspective is given from a professional standpoint in which honesty is variously assumed, supposed and abused.

What should a scientist do when his life is in professional danger and not just his own life, but those of his staff and of the general public? What should he do when he realizes that the problem is not specific to his locality but generic around the world? Should he blow the whistle when he knows he has no funds for safety - when he knows that his laboratory accounts are grossly misleading and that he has no power to correct them? Should he rat when he knows that radiation rates, hundreds of times above legal limits, have been emitted? Should he do this when he knows that the political instincts of the judiciary are uncertain and that universities have powerful resources to bring against adversaries? What he actually does is to ensure that he can continue to profess the truth from secure ground - somewhere that can allow this by preventing his livelihood from being robbed from him and somewhere that has a more independent legal system. He relies on survival instincts at both biological and institutional levels, instincts that combine with the best reasoning he can bring to bear. Truth, honesty and credibility are what his personal career had been based on and they have the power to bear his undoing. As much power as undid David Kelly. Scientists are not able to expose the callous errors made by unscrupulous managements.

Academics are trained to see many sides to a problem. When he sails between Scylla and Charybdis, the scientist needs to know it. Formal complaint will make it impossible to fund his laboratory with vital supplies and equipment. Shrill warnings will jeopardize the public as much as silence, on the other side, will risk the health of all. The best he can do is to complain from a distance and refocus his energy where his science takes him. Science is his bedrock. He moves his foundation from a beach of shifting sands to the firmness of a foreign island. When they persecute in one city he moves on to another.

Time was on his side. He could instruct his staff and make his own exit before the local administration would realize in what danger they were. There are hazards everywhere. All writing is a kind of fiction and this is a story of what happens beyond press gaze. Familiar, warm and cozy categories are not what his story is about. It is about seeing and saying, as well as possible and for the good of all, what is out there. The tale is about drawing the most from mundane lies.

Science is dangerous and NASA is witness to it. For the wider community, there are ways of reducing the damage done by run-away science. One method of inverted hierarchy is described. This method has been tried, but on its own proves insufficient. Something more is needed, not just integrity but something discriminating. The Law of Universal Mendacity shows what that is.

Other scientists have preached doomsday. In his book, *The Final Hour*, the Astronomer Royal opines a fifty per cent probability that our present civilization on earth will not survive the twenty first century. Considering various catastrophic threats, both natural and man made, he bet that bio-terror will kill a million people before the year

2020. His view is, however, in some ways optimistic and in others descriptive. Among the millions of galaxies out there are billions of planets like our own. Yet we have not observed life there, while "they" have not, apparently, discovered us - in spite of "them" having, on probability, much longer for discovery than the fifty years of modern communications that we have owned. Is the reason for the glaring omission that intelligence is self-destructive? We add an inherently unsafe scientific environment and, without objecting in other ways to Rees' speculative warnings, we add a warning over gullibility in the choices we make.

About end-notes: there are not many. The general purpose is to reference published work where a wider perspective is wanting. A few end-notes make side comments that, though maybe helpful, would cause a distraction in the text. More generally, the information, where it is not referenced, should be taken to be first hand, whether it is from the direct experience of a practicing scientist or from his general observation. For geeks, more complete sets of references can be found in the academic tomes cited. There is no need to copy their lists here.

About terminology: the words 'assertion' and 'statement' are not interchangeable. `Assertion` carries a stronger willful element and so arouses weaker feelings when the implications are examined as in this book. As is explained below, when we 'state' propositions we use a process of abstraction.

Finally, this book invites assertive criticism. When the critics have gone, the book will stay. Beyond Universal Mendacity rises a Law of Universal Perversity. When our critics intend to admonish they will in reality be affirming. Either way, nothing much will change. Our naive hope is that they will, on reading this book, rest more easily during their short stay in this world, as we do for writing it. Perhaps even, our final hour and the end of civilization might be postponed.

Bo De Yang

Chapter 1

PROBLEMS WITH PROBLEMS

Can you force *anyone* to be honest? You can't force *everyone* to be honest and it is difficult to persuade them; but we can always avoid being dupes. Common sense? A few words of warning: the unfamiliar may accompany feelings of skepticism but you may also recognize an elemental truth. The reader is invited to exercise skepticism and that is even recommended as proof of what is written. Common sense is a poor guide, and if you become uncomfortable, you would do well to analyze exactly what it is that bothers you. The opposites that are intended and experienced in assertions are not in fact contradictory, and the reasons why are illustrated and explained.

Like the philosopher Bernard Williams, we believe in truth and honesty and we do so because credibility is our aim; but perhaps unlike him, we do not expect to find it, not even professionally - even though our profession is different from his. Our stance is descriptive and reflective. In life, it is usually not possible to explain to others the complexities of our arrangements or beliefs, and we expect others to suffer the same limitations as we do. We adapt to informants - as we must when speaking to children, for example. More generally, the recognition of a need sometimes to communicate ideas and situations - at whatever level of complexity - in fact imposes moral imperatives. Philosophers who need credibility from critical audiences had better be right and honest if ridicule is to be avoided; but for the wider perspectives of life, of knowledge, and of religion, there is a complementary need for understanding what is being offered in the form of assertions. If I want to be believed here, I had better say it honestly; but I am also reflecting on the limitations of this approach for the philosophical parts of normal man or woman in the wider perspective.

We shall illustrate how, from the Law of Universal Mendacity, propositional calculi are abstracted by a process of closure. The abstraction of geometry creates propositions that are simply true or simply false. Operating beyond this level of abstraction, the Law of Universal Mendacity is not a physical law; but it results from the way, in nature, we are informed about states of affairs. Its model is not the knowledge of geometrical theorems based on axioms, whether self-evident or contrived; rather the opposite as will become evident.

The book attempts a state of mind that will help in answering several questions experienced by thinking people of our time. What can you believe? What is a lie? Are white lies reprehensible? Can honesty survive? Is cynicism the trait of the fittest? Is morality immature or out-of-date? These questions underlie our great trauma - one that continues to subsume many of the best movie stories. How did the Second World War occur? How were so many deceived? How did allegedly civilized people do such barbaric things to each other? Has anything changed? Because of space and time constraints I offer only a constructive attitude; the next step is yours.

My first book was in a specialized scientific field. Science is, in most people's minds, the area where knowledge and truth are most conclusively pursued and agreed. Philosophers, even if they rarely study it, have, since its successes in the seventeenth century, taken science to be the paradigm for knowledge. I have tried to show how the Law of Universal Mendacity applies in this context, and from there it will be relatively easy to see how it applies to other intellectual disciplines.

There are serious problems in discussing the basics of truth and falsity and they have been discussed before many times. We can therefore skip over well-known criteria of factual correspondence, consistency of argument and rules of evidence. We also skip over the classical theories of knowledge, expounded for example in Plato's *Theaetetus*, that have been assimilated over ages into our common sense. More profoundly, a group of problems result from asking "What is the basis in truth for the assertion made about it?" They are illustrated by various paradoxes. First an easy one: A barber shaves all persons who do not shave themselves. Who shaves the barber? No problem, *she* doesn't shave. But what do you believe from the Cretan who said that all Cretans are liars, or from this one: the philosopher who said he didn't know anything, betrays the "knowledge" that he doesn't know anything? Can you take him seriously? One method that can be used to overcome this type of self-defeating difficulty is the exercise of power or authority which might take this form: "I belong to an influential university department with good connections to editors and we make sure that what we say goes." Williams overcomes the obvious shortcoming in this approach by a method of "Genealogy" derived from Nietzsche and Hume. It is a method that makes an evolutionary case for a bias towards honesty, sincerity and authenticity that arises from a human need to communicate ideas. In fact, of course, nature is full of deceit - in predators especially; it selects camouflage in prey.

As an example up the evolutionary tree, consider behavior in poker players. These display, individually, the need *not* to communicate what one knows, and they display, moreover, a need to communicate something opposite, so it may very well be that deceit has been selected before honesty. In some environments honesty may seem more prevalent, in others deceit and your view may depend on the type of upbringing you received. But anyone's view should also be influenced by the different degrees of communication that various circumstances demand. In real life, a 'too honest' appraisal can be twisted, all too often, to support a view or intention opposite to the appraiser's, and especially so when speech is careless. We apply an evolutionary approach to moral problems because, in heaven or on earth, the benefit of being good is limited if goodness doesn't survive here?

Is there really a bias towards truthfulness? Consider a cosmological argument against this. Beyond our solar system there are millions of galaxies and billions of planets. Some must be like our own. If there is not life out there then *we* are extraordinarily lucky to be alive, which would have to be improbable. If there is life out there, it is probable that there are civilizations that are hundreds, thousands, millions, or even billions of years ahead of us. Yet we have not detected them, and they do not seem to have detected us. A probable explanation is that civilization is self-destructive. From the history of the twentieth century on earth, it is easy to see how. A suggested explanation is that deceit is a contributing cause. If so, the bias is towards deceit.

Like Williams we acknowledge the need, in survival, to communicate ideas; but we concentrate on assertions and discriminate true logical propositions in statements. What we write is consistent with our writing of it. Truth is not, for us, simple as in logic. Our cave is different from Plato's. He thought that he could see clearly only with philosophy. Both caves are initially dark; but ours is endless and we examine it with a flashlight of words. The television lights never switch on. Often things and affairs are not as they appear. Monsters look like rocking horses and *vice versa.* We say what we can and listen to what others say with the purpose of moving forward. Life is short, but we take as little as we can at face value if there is one. Our book is shorter only because others have been longer.

This is not an academic exercise; it is written lightly with a heavy warning. Nor is it a rehashed thesis or lecture course; but if you find the second chapter hard going, notice that others will find this the part worth reading. It is denser than the remainder of the book. Don't give up, but skip over it and return to the chapter later. You will find, most likely, that the subsequent examples and amplification will make that chapter easier to comprehend. Remember that a seed must die before it sprouts and yields its harvest.

Bo De Yang

CHAPTER 2

THE LAW OF UNIVERSAL MENDACITY

Words contain real meanings that oppose their intended meanings. The Law reflects back on itself with a double opposite: *in opposing reality, words have intended meanings.* You might have to think them through, but rightly understood, these assertions are not contradictory. In the fact of assertion, *the intended meaning is one part of the whole.* Strictly, the Law applies to assertions and not to statements of propositions. As we understand them, propositions are either true or false; but there is always some truth in the opposite of an assertion and often the context supplies a stronger and contrary truth than was intended. When you read Universal Mendacity, capitalized, we are referring to this Law.

Did you ask what is meant by "intended"? Why did you ask? Your intention is your reason for speaking or asserting.

You can review where we're headed by scanning the contents page. In a later chapter we'll consider the paradigm of knowledge, in science, and we'll elaborate subsequently with further examples and implications.

9

2.1 *Examples from public and personal life*

Nowhere are the contradictions inherent in assertions more celebrated than in public life. When the president says "We'll get 'em," you note his determination but ask, "When? Where? Can he? Ever? And the weapons of mass destruction?" Many political promises, perhaps even most, have been broken because they were too wishful - for speakers and listeners - whether well or badly intended. Back sixty five years, "Peace for our time" meant, in consequence, exactly the opposite. The *real* meaning came from the context: the memories of a prior and devastating war, the policy of appeasement, the inability of the prime minister to impose his will, and the encouragement for belligerent dictators. Yet these assertions were charged with meaning and there is a real meaning which supplies that charge and began in opposition to the assertion made. The real meaning is the root of these statements' significance.

You can work out from political denials that this inherent contradiction occurs, and even invariably occurs: context is the most

common excuse given for political *faux
pas*. Context is often contentious and for
various reasons. Sometimes
commentators deliberately take words out
of context in order to misrepresent and
discredit; at other times, with more
perception, the context seems to reveal
underlying reality. Spoken words may
sometimes be inadvertent and may
furthermore be the repressed outcome of a
Freudian slip. Less perversely, speakers
are often not sufficiently aware - until too
late - of the subtleties of context and have
to subsequently clarify or cover up. For
the savvy, the clarification can be
intentional and used to make a further
point, like a supplementary question in
parliament. The context is part of the
meaning. All of this is sufficiently
obvious to make further examples
laborious.

The Law, or Universal Mendacity, is not
restricted to public life, but for the same
underlying reasons applies universally to
assertions. True lovers do not need to
declare their love. When children demand
declarations of love, you try to work out
why, and what is wrong or missing.
Nowhere are declarations more

11

forthcoming than in religious discourse. "Lord I believe, help my unbelief," discloses the tension pulling the will between doubt and recognition. Whatever is intended, the context poses the opposite. An assertion is inane where this is not so. It would be like a philosophical illustration of the "all cows eat grass" type or a "simple truth", so simple as to have no applicable meaning.

At the other extreme, assertions can be made so complex as to fuse into primeval and meaningless verbiage - assertions so bland, hedged and convoluted that they come to mean nothing. Simplicities are needed along with the complexities and these too have a context. Compare, "pass the salt" with "salt, in a manner of speaking, has a sort of taste that may slightly alter, favorably or unfavorably, the flavor of things you might eat, depending on your peculiar sense or preference for tastes of a one kind or another, if you see what I mean - not all salt, obviously, but some salt and then only if you take it with a pinch." No liability there.

With the inherent opposition found in assertions, the triad of truth, honesty and morality have to be understood in their context and complexity. *Honesty is the intention to fully communicate the truth as perceived.* By "fully" we mean adequate to the demands or interest in the subject. Is absolute honesty possible, and if possible, would it be moral? Consider a few examples.

2.2   *Morality of truthfulness*

Morality begins with childhood. Children become confused if they are not taught simple truths. They are frequently deceived; but as often reminded to tell the truth themselves. This makes it easier for providers to care for them and manipulate them. You'll most likely remember this from your own childhood and even from your experience of parenthood if you are so blessed. The constraint in rearing is one explanation for the bias towards honesty professed in Williams' Genealogical method. Mental conditioning in children is often not properly criticized and unlearned in mature life.

The growing child, especially if it has been well brought-up, will make an effort to modify the simple truths, learned in

13

youth. Most often it will learn what philosophers know: that good actions have desirable consequences; that people act for what they perceive to be best; that it will not be able to make an election with every step, but it will grow to rely on intuitions developed at critical moments. Furthermore, it will learn to make assertions carefully if it is to avoid the consequences of ill-considered recklessness.

If it wants to be believed, the growing child will need to build credibility. It will pick a route between well-intentioned dreamer and effective person of affairs. It will be penalized for breaking rules on lying while using and extending those same rules for its own purposes, whether immediate or remote. It will certainly not believe everything it is told. It will learn to discriminate innate signs of bad intentions. People telling deliberate falsehoods, for whatever reason, tend to sweat or look sideways. The eyes are more difficult to deceive than the mouth. There may be other ways and manners, too, of telling a lie. Lie detectors don't always work; but the telling of lies, generally leads to a loss of credibility -

especially in people with long memories or when there are opponents at hand to expose or ridicule the lie. Because of Universal Mendacity, denial makes weak opposition; ridicule is more effective, especially when it reveals the underlying reality. If the growing child needs to be believed, it will - unless the audience is naively and wishfully credulous - most likely tell the truth.

2.3   *Deliberate deception*
Even deliberate deception depends on an appearance of truth. But when the deception is doubted or revealed it says as much about the deceiver as the deceived. It is odd but true that all assertions imply, as in the wishful assertions of presidents and prime ministers - their opposite. In context, Washington always sort of lied at least when he made assertions; simple propositions are a different matter as we'll see. But first, what is a lie? The lie resides in the intention that betrays reality. The great writer of dictionaries, Dr. Johnson, would call any untrue assertion a lie. Sometimes he would distinguish by saying "He LIES and he knows he LIES." His usage of the term was empirical enough, actions being more easily known

than intentions. In such a case the intention belonged to the person making the assertion. For 'lie' we choose the second meaning, as Johnson embellished it, because that is the more commonly understood meaning and because of the importance of intention in Universal Mendacity. In the sense of "knows he lies," maybe George Washington never told a lie after all. There is a sociological literature on lying. We distill a lie to be *an assertion intended to deceive a dupe about affairs that include the intentions and state of mind of the liar.* Even so, wherever the intention is to be found, whether in the writer, or in the reader or even in the words themselves, opposite real meanings occur, and always occur, in the assertions pronounced. Universal Mendacity makes even the reality of a lie reveal truth through its context.

2.4 *Certainty*

If the child will not entirely believe others, will he believe himself? If there is a real meaning for its words that is opposite to the child's intended meaning what purpose can there be in speaking? Proceeding further, if there is an inherent opposition in the making of meaningful assertions, can the growing child be sure of anything, whether taught or learned? All thoughtful children attempt the method of universal doubt at some point in adolescence. It invariably stumbles on Descartes, who discovered certainty in the observation, "I think therefore I am." But what of a child who knows that any intended assertion relies, for its meaning, on an opposing reality? Can it find certainty in anything? Will it be able to say, "I think therefore I doubt that I am?"

Maybe and for several possible reasons. If asked how it could say this, it might give various answers ranging from the mundane to the irresistible; but they all confirm the Law. The child doesn't have to be serious but let's suppose it is. (Actually, an unserious assertion would be a pretense that opposes its intention - the speaker's or the listener's - to the reality.) The child could contradict the *Cogito* to question the meanings of the words, especially "I." Is this the immortal soul, or the center of consciousness, or a permanent unity that lasts a lifetime, or the momentary actor strutting on a stage, or his own peculiar smell, or any of many other meanings that can be given for the

17

first person pronoun? After a failure of memory, for example, I may not exist in one of those senses. Or moreover, if "I" were nothing, then "I am" wouldn't mean what was intended. There is a common figure of speech: *I am not myself.*

But to come to the point in the *Cogito*. Is there nothing about which he cannot be certain - nothing that depends on an uncertain meaning? The child could analyze this for a long time and would only reach a conclusion when he ran out of time. There is a research industry in this type of analysis. Something, it may be, happens in the asking; but its reason for asking remains, as a contextual fact, open to unending examination. Something happens mentally when it utters the *Cogito;* but what exactly that is, remains indefinitely open. Consider next another area in which certainty has, since the Ancient Greeks, been supposed.

2.5   *Types in contradiction*
The two arithmetic propositions 2+2=4 and 2+2=1 are contradictory and the second is a contradiction in terms, unless it supposes counting modulo 3. These are propositions of the same type. The

intention in a proposition is restricted, in this example, by the axiomatic and logical system assumed. *In this restriction, the arithmetic propositions, or statements, differ from assertions.* Assertions cannot be directly compared in this simple way. Think of the 18th century debate between idealists and realists. Bishop Berkeley, taught that the *esse* of material things is *percipi.* A supporter had this from Dr. Johnson, "Pray, sir, don't leave us, for we may perhaps forget to think of you, and then you will cease to exist." Far away and years later, Hegel synthesized, "In the idea of the will the real and ideal are united." This is the same opposition of real to ideal that we find in Universal Mendacity; only we apply it dialectically to assertions as human actions. All assertions are human actions.

In fact, under Universal Mendacity, the real and ideal started separately and crossed without synthesizing. Something happens: they knock together and then go their separate ways again. Mind will construct another fairy castle.

At the conclusion of this book, the assertion of Universal Mendacity can be taken as fact. A fact *is a proposition that is accepted as true. Facts are representations of reality.* Two propositions are contradictory if they cannot both occur at the same place and time. For example: "I am writing this" contradicts "I am not writing this." The words seem to have meanings; but no purpose or intention is ascribed besides the illustration of a formal contradiction in "closed propositions." To illustrate what closed propositions are, consider first completeness in axiomatic systems.

2.6　　*Consistent or complete axiomatic systems*
Inconsistency is typical in axiomatic systems. "Some classes are members of themselves and some are not. The class of not-teapots is a not-teapot. Consider the class of all classes not members of themselves. If it is a member of itself it is not; and if it is not, it is". Russell's theory of types responded to his paradox: the class of all classes is of a different type from the classes themselves and so modified logical rules apply. Godel went further: by assigning numbers to statements of an axiomatic system and using the property that the number of prime numbers is infinite, he proved that if

an axiomatic system is complete, it is not consistent; and if it is consistent, the system is not complete. That is, either a statement can always be made which is beyond the axiomatic system so that the system is not complete; or the system contains an inconsistency of the sort a=b AND a≠b, as illustrated by the class of all classes. Universal Mendacity is not a theory of types, but its assertions contain opposed meanings: the ideal intention opposes the real context, and this complexity is what provides a particular assertion with its meaning. Mathematics is almost an exception to the Law. If mathematics is taken as consistent, it contains "closed propositions", but it is not then complete.

2.7 *Did Washington ever lie?*
So when Washington never told a lie, he was revealing real meanings in opposition to his intended meanings. Cynics know this. When they are unconvincing, their cynicism is too general. To be convincing, you have to look deeply into that opposition. Plain speaking, whether in public life or in opposition, is not enough. The simple doves become more credible

when they are also wise as serpents. This is how you become so.

When you hear an assertion, look to the opposite because you'll need it to find the full truth. The truth is the mental and verbal reflection on reality, the state of affairs. It is not that there may not be something credible in the assertion itself; but to understand the assertion you'll need to be aware of the reality that provided it with meaning. "I cannot tell a lie, I didn't do it," was believed in the appropriate manner. The ambiguity in the assertion betrayed a cover up on its own, even had there been no supporting evidence.

2.8   *What is intended meaning?*

The previous paragraph refers to assertions, not propositions. Propositions are either true or false, like the two arithmetic equations given earlier as examples. They can be stated with absolute honesty. *Assertions add intention.* Intentions are often difficult to determine. We think we know *our own* intentions, but we are sometimes mistaken ourselves in the assurance, owing to subconscious influences. Determining the intentions *of others* is sufficiently difficult that it is often wiser to stay close to descriptive fact. When words are spoken, the speaker supplies an intention in speaking and the listeners supply their own intentions. Sometimes communication and agreement occurs. The words themselves are do not simply describe facts, but carry various cultural meanings in conditioned, ambivalent circumstances.

Assertions are less like the propositions of mathematics than they are like the claims of confidence tricksters or conjurers. A trick depends on seeming credible. Assertions describe reality, but they never give the full picture that is photographic in detail, three-dimensional, tangled with emotions and plots, and dynamic. Assertions are a yacht, leaning away from the wind, the hull cutting water on both sides, not necessarily tipping over, but moving forward. Sometimes the context shows that, for the tack taken, the sail is on the wrong side and the wind blows from the wrong direction. This is when we think of mistakes and downright lying - a deliberate miss-statement of a state of affairs intended to deceive. Under Mendacity, absolute honesty is possible in stating propositions; but not in making assertions.

Universal Mendacity is as much a reference back to absolute idealism as it is a progression from the logical positivism of the nineteenth and twentieth centuries: "the meaning of a proposition was its method of verification." But the method was disembodied. It applies rather to propositions in closed logical systems than to assertions. These are made and interpreted by people with purposes and intentions. According to Popper, the scientific method operates by disproving theories by contradictory experimental evidence. Experiments are designed for this purpose. The *Logic of Scientific Discovery* is open ended like Universal Mendacity. As we know, an assertion has significance if, and only if, there is a reality opposing the intended meaning. For significance, the words imply a real meaning which stands opposed to their intended meaning.

2.9   *Closure in practice*
Universal Mendacity is a law of underlying uncertainty. You can imagine Dr. Johnson: "Pray, sir, say the opposite of

what you mean so that we may know the truth." That would be to mix the intended and the real. The truth is not so simple. In Universal Mendacity, if the real opposite and the ideal intention are interchanged, the yacht changes tack. Humor is frequently based on the ensuing irony: 20th century Europe was civilized; the 21st century United States is peace loving. However there is a practical way of reason and that is where we join hands with common sense.

We sometimes treat assertions as if they

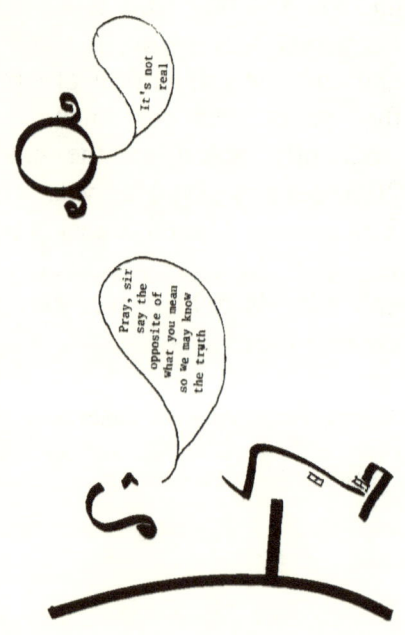

were propositions. We restrict them by apparent closure so that we can move our attention forward. A proposition is closed when its meaning is derived axiomatically, as in mathematics. It is either true or false, and the opposite is correspondingly false or true. As a further example, closure occurs whenever a new scientific discipline is based on an earlier one, like molecular biology on the structural analysis of DNA. *Closure converts an assertion into a proposition.* We simplify and abstract.

Science is practical. Widely accepted laws can be regarded as approximately closed even though they remain open in the final analysis. Some examples are given in the next chapter. How intention is simplified or discarded is discussed there. Nowhere is so much agreement supposed to occur as in science, the model of knowledge. But *The Structure of Scientific Revolutions* complicates a naive theory of objective and dispassionate discovery in science. Kuhn teaches that there are paradigm shifts at critical moments in the progress of science. These too, leave space for opposite meanings and some examples will be given.

27

> Like science, history and the other disciplines imply a fine line between closed propositions that are accepted for the sake of argument, and open assertions under critical review that are subject to Mendacity. This will become clearer after the following chapter.

It is evident that we appeal to no "Genealogy" for justification; but to a more primitive consistency, born of a kind of agnosticism loosely married to abstract systems of closed propositions. *We are of course writing about the meaning of meaning. What we mean is that words carry not only the intention of the speaker or hearer, but conspire with them to reveal a situation of significance that is fuller than the simplicities that intentional language can represent.*

CHAPTER 3

LAWS OF SCIENCE - ASSERTIONS OR PROPOSITIONS?

Simple propositions are either true or false. They can be represented on truth tables, following the axiomatic rules of logic. Opposed to a true proposition is a false proposition. Assertions are not so simple. Their logic is a dialectic of opposites.

Scientific laws are sometimes taken to be not only propositions, but true propositions. The laws are thought to be true, not as mathematical tautologies (all unmarried men are bachelors), but synthetically true (Hume was a bachelor), following experimental evidence. Scientific laws, like the inverse square law of gravity, seem to be as real as kicking Dr. Johnson's table. This apparent reality is complicated when consideration is given to the actual discoveries of these laws. The fact that some apparent discoveries have been frequently disputed, as much before the time of Galileo as since then, is due to many reasons or circumstances, including bigotry and special interest. But even without those unnecessary, narrow and negative features, the progress of scientific theories normally

29

involves disputes and these are sometimes bitter and prolonged. They occur in the context of intuitively held beliefs of individual members in scientific communities, even when regarded as unbiased. The disputes are sometimes not resolved, and the resulting antinomies can last for centuries. More generally, there invariably exists a period of doubt before a law is widely accepted as true. This doubt occurs during testing of a theory whether by predictive experiment or for consistency. Since a scientific law is a cultural phenomenon, it inevitably includes intention and, following Universal Mendacity, implies for its significance, an opposite real meaning. Scientific laws are therefore a testing ground for Universal Mendacity.

A modern example of the disputes that occur in the formation of new science is described by Joao Magueijo. He invented a speculative theory to explain several cosmological anomalies. This Variable Speed of Light theory overturns a simple understanding owned by some physicists, because in Einstein's Special theory of Relativity that Speed was constant, determined as it is by "physical laws that are invariant in all inertial (static or moving without acceleration) reference systems." The Special theory arose from a remarkable observation and the theory preceded the General theory, invented also by Einstein ten

years after the Special theory. The observation
was made in the Michelson Morley experiment.
This was an experiment to measure the speed of
light in two directions at mutual right angles.
What might you expect? Suppose that light flows
in an ether like a boat in a flowing river. If the
propeller runs at constant speed, then, relative to
the river banks, the boat runs faster downstream,
slower upstream and at some intermediate speed
across the stream. The Michelson Morley
experiment showed that the speed of light is the
same in all directions and in particular is
independent of the rotation of the earth. There is
no ether. It is strange but true that the speed of
light is independent of the speed of the light
source and of the speed of the observer. If you
think of an F18, flying at an earthly velocity of f
miles per hour, firing a missile with speed m
relative to the airplane at a ship moving front-on
and traveling at speed s. Then an observer on the
ground will measure the speed of the missile as
$f+m$, and an observer on the ship will measure the
missile speed as $f+m+s$. These three
measurements are relative and different. But if
the three observers were to measure the speed of
the light that they can see emitted from the
traveling missile, they will all measure the same
value, c, the *speed of light*. This value is
independent of the relative motions of the emitter
or of either of the observers. So long as they are

inertial, c depends only on physical laws. By contrast, the *wavelength of the light* is relative: it looks redder to the pilot and bluer to the ship's captain, compared with the color seen by a stationary observer on earth. In Special Relativity c is absolute; space and time are relative. A well-known consequence of the theory is the equivalence of energy $E=mc^2$ with mass. It turns out that the energy release in the fusion of so small a mass of hydrogen as one gram is greater than the chemical energy released in burning twenty thousand tons of coal.

In the General theory of Relativity, space is only locally inertial but the speed of light remains constant. When you want to know if an edge is straight, you eye it down the line because you know that light travels in straight lines. You can think, if you like as Newton did, that space can be mapped onto a Cartesian grid. Some time before Einstein, the philosopher-scientist, Ernst Mach, thought space should not, in fact, be so. Space should depend instead on the things inside it, and this is what General Relativity explained mathematically. In consequence, light that passes through strong gravitational fields, for example near the sun (it can be seen during a total eclipse) bends in our perception. Patterns of stars close behind it are then deformed as compared with the normal pattern you might see in the night sky.

In a Variable Speed of Light theory, *c* is allowed to vary and this has special solutions, for example at critical times after the big bang or in the vicinity of black holes. However the theory is so far unproven, has had a difficult birth, and remains contentious. Not all theories are as contentious and the following example is milder. Einstein had, in his General Theory, added the cosmological constant to his geometrodynamic law now represented, comparatively simply, as: $G=8\pi T$ (the curvature tensor is proportional to the mass-energy tensor). He made the addition because he thought the universe was static. When Hubble showed, soon after, that the universe was in fact expanding, Einstein called the added constant "The biggest blunder of my life." Though the constant is known to be wrong as originally written, it continues to serve usefully in current speculation.

History holds many stories of novel science with contentious beginnings. Another example shows, firstly, that scientific laws are assertions, and secondly, that they embody real opposites. The corpuscular theory of light propagation was supported by Descartes, Boyle and Newton. On this theory, light moves in straight lines. The theory seemed, over a long period, to contradict the wave theory. This latter theory was strongly reinforced by later experimenters and theoreticians including Huygens and Maxwell, but conflicted with early 20th century developments that included the phenomena of black body radiation and Planck's law of quantization. The phenomenon of diffraction of waves requires the bending of light. If you look at a street lamp through an umbrella at night, you'll see a pattern of dots because of diffraction. The pattern anticorrelates with the mask of holes in the fabric weave because umbrellas with finer weaves show broader patterns. Wave interference is responsible, and the reason in physics text books is not hard to follow.

The antinomies of corpuscular and wave behavior were inherited by quantum mechanics where the probability amplitude was discovered to predict the chance of detecting individual events. The probability amplitudes for occurrences of such events are determined by wave fields; though the individual events are not independently predictable. There was an extended debate on the foundations of quantum mechanics, Einstein being both a founder and skeptic.

The antinomies are illustrated by Schroedinger's cat. He proposed a well known thought experiment in which a cat was placed in an enclosed box to be killed by some mechanism triggered by the radioactive decay of an atom. Though the *probability* of the cat being alive at any moment after the start of the experiment was exactly known, the only way it could be known whether the cat was alive or dead was by lifting the lid. This illustration is seventy years old, but we still do not know any other way of discovering whether the "cat," as a quantum mechanical event, is "alive" or "dead," since that is a matter for undetermined chance. How can it be that the probability for experimental outcomes is completely determined while an individual event is otherwise entirely unpredictable? That is, the continuous probability wave field is determined but individual quantized events are not. For practical purposes and for reasons of closure, most scientists accept the unpredictability as a fact, though there have been dissenting voices since the earliest days of quantum mechanics. Many theories of quantum mechanics are supported separately by the beliefs of the theorists or experimenters upholding them. The various beliefs that underlie such fundamental theory show that scientific theories and laws are assertions, not propositions. Quantization is well established for electromagnetic and nuclear forces, but even now, the quantization of gravitational fields is undemonstrated and unresolved: if there is a smallest unit of gravity, it has not been measured.

The wave and corpuscular theories provide mutual opposites, both of which are real. In the same way, the idea of a determinate, quantized, probability amplitude that controls an undetermined particular event, derives its significance from the mutual opposites.

For any theory to have meaning there must be a counter-theory. Experimenters, when they are not engaged in exploring new territory, follow Popper in trying to produce evidence that proves either one theory wrong, or proves a prediction to be true. Experimental results are sometimes questioned, sometimes reinterpreted. They may be more or less successful in proving the claims associated with them. For practical reasons scientists accept a body of knowledge and direct their attentions at new areas for examination; but in fact no law of nature is unquestionable, from relativity to space, time and gravity. If there is a cultural awareness about what is progressive; there is a corollary about what is sterile or iconoclastic. The awareness produces virtual closure and this closure drives forward the mainstream of scientific development. For this reason, most physicists are mechanics.

Generally therefore, physicists can, and do, treat some of their theories like systems of logical propositions having axioms and necessary conclusions. The closure can be happily executed with accepted and undisputed theories. Statistical mechanics is an example - the theory which applies most probable outcomes for the macroscopic properties of heat, volume and pressure that arise from the random motions of large numbers of atoms or molecules (whether deterministic or not). But for more disputed theories, doubt implies real opposites to intended assertions.

As systems of logical propositions, undisputed theories lack assertive intention and are not subject to Universal Mendacity. The Law applies to assertions. It begins to apply where those theories are applied to research and to the development of new theories. It applies more strongly to the building of scientific laboratories. Scientists occupy positions of special trust in the public interest. They expend large sums in defense of its liberties and in progressing its technologies, communications and culture. Often they engage in the education of the young. Yet they are, for mandarin reasons, their own collective judges about how those funds are spent. If the complexities of scientific politics are hardly distinguishable from national politics, this gives the key to understanding the management of science. The careers of individual scientists depend generally on getting access to those funds and even on controlling them. However, there is nothing in the moral structure or behavior of these same scientists that gives ground for believing that the funds are spent disinterestedly for the public good - any more than such funds would be spent in any other part of civil life. Often the management of funds is incompetent and worse.

That is why large laboratories are sometimes built at great public expense after the founding notion and reason for a laboratory has been discarded and the edifice, far from having a justification, has become a public threat and health hazard. Sometimes these laboratories emit life threatening radiation on the public - radiation that is not controlled for various reasons: lack of resource, lack of awareness, ignorance, special interest and so on. The threat is sometimes disguised by a cloak of bureaucratic deceit and conceit. Scientists have the same vices as corporate America. The more scientists claim to be incorrupt, the more suspicious - another consequence of Universal Mendacity.

Here is an anecdotal example of the way the waste occurs. Government has to be seen to act and proposes an initiative. Staff are hired, sometimes from abroad. Universities and Institutes may resist but are obliged to comply with higher authority. The new staff propose initiatives. A case is made. International conferences are organized. Promises of support arrive. A substantial grant from public funds materializes. Laboratories are built and equipment ordered. There are rumblings about the selection of equipment. The support from industry changes owing to government inspired changes in management. The support is withdrawn.

What happens next? Decisions are taken by various interested groups, each of which is incapable of independently evaluating the scientific merit. As it turns out, the equipment ordered for one purpose is not in fact useful for another, though a pretense of continued usefulness becomes a line of least resistance. Nobody stops the freight train. The laboratory moves from white elephant to dead duck, but a great deal of further expenditure is needed before it transmogrifies to acknowledged dodo and contaminated junk that is too dangerous for mere scrap. For a long while, nobody is willing or able to make a decision. *Those who know, don't fight.* Knowledgeable experts resign and are replaced by consoling fantasists. The new *"experts" are hired for their level of incompetence.* Accounts are bungled to mislead. Safety and radiation are scattered to the winds. The original initiative has turned into a public hazard with no extant justification nor any mitigating factor.

At this point, nobody is responsible and the longer the project continues, the more established and easier becomes the cover-up; but the waste of public resources continues along with radiative exposure. The administrators, who owe their positions to their concentration on scientific politics, persevere in willful ignorance. They imagine that by hiring experts they became competent themselves, and even more so after the experts departed in frustration. Scapegoats, whose prior warnings had been wilfully ignored, find other places that are more conducive to breakthrough science. Back in the lab., there are no prizes, no results and many wasted watermarks. Two kinds emerge: those that concentrate on politics stay; those that create leave. The ridicule is distant and muted but profound. The danger is real.

From this chaos an echo of Parkinson's Law reverberates. *Administrators rise to their levels of ignorance.* We call this executive rule 1.

What else can be done? How can the public be protected? Following Universal Mendacity, there is little value in asserting a particular and irregular occurrence. The assertion will betray itself owing to the application of the same Law. Pseudo-martyrdom in established courts, against massive opponents, is generally ineffective. No courts are free of human bias and interest, in the sense described in Hart's *Concept of Law*. Judges have careers to pursue. With less formalism, whistle-blowers take large risks and often fail to survive. Under Mendacity; serious reform falls to those with talents for cartoons or comedy, however ineffective they are taken to be.

If the waste of equipment doesn't occur in laboratories all over the world you should wonder why not. One institute bought, in error and at considerable taxpayer expense, the best commercial electron microscope and built a purpose-designed laboratory as specified by the manufacturer. But when the microscope was delivered, it was relocated to a room next to transformers. So the world-class magnifier of atoms became fuzzy translucent. Unfortunately, high-energy electron microscopes are also radiating sources. Administrations are not threatened by failures in safety and loss of life. In NASA, following repeated accidents, an inquiry was held into the loss of the Columbia space shuttle and a reform of safety procedures was required. But as is typical, the inquiry took place *after* the accident. On a broader scope, we should expect the same behavior with regard to the survival of humanity.

In an age of science, what has not been invented is a social structure with the required antidote of competent, inspired and purposeful creativity. The best technology is rarely implemented. Product, in science, is lost in its haze.

Three principal concerns arise: one is the credibility of science; another is the management of science with public funds and the third is the impossibility of providing public safety. Beneath these concerns lie the same underlying principles: *people will do what appears to them to be best* (executive rule 2), and *careers in science depend on unenlightened self-interest* (executive rule 3). We'll return to the three executive rules in the next chapter. Before that, consider the following objections to what has been written: firstly a supposition of academic honesty and secondly consequences of the repeatability of physical experiments.

The ivory towers are supposedly biased towards truth and truthfulness (aren'they?) for the reason that the competition existing among scientists results in a kind of self regulation. A charge of academic misconduct would normally be brought against overt lying - wouldn't it?

Actually the opposite happens and for reasons that lie in the sociology of science. Consider its basis in academic tenure. This is the system by which, in many universities, staff are dismissed after hiring if they do not meet the approval of tenured staff. Even staff who have held tenure elsewhere can fail in the process through the manipulation of colleagues. An examination of opposites in the context of how the academic system works, shows that the supposition of truthfulness is naive. Firstly, academic positions are much sought after and generally earn more than a national median. Academics with talent do not need tenure, and often resign it; while those without talent have to struggle for the tenure as their key to personal economic development. The acquisition of tenure becomes, therefore, an exercise in manipulation, sycophancy and dishonesty. The community that is supposed to regulate the conduct of scientific discovery is itself based on trickery. That is why university departments have, over a long period, rejected able staff that have thrived elsewhere and sometimes earned the highest prizes. The injustice is actually only an inconvenience but the humor is on the Universities. In Mrs. Thatcher's Britain, academic tenure was weakened. A generation of "refuseniks" kept their contractual tenure at the expense of promotion. Tenure is counter-productive. Redundant academics are relieved; recruits are hired for their ability.

A consequence of the competitive bias, in science, away from honesty, when compared with the norm in a population, affects the credibility of scientific theory and report. Given the importance of publications for the careers of all scientists, why should occasional press reports of scientific misconduct be more than the tip of the iceberg? Short of the extreme of fraud, there exists, moreover, a history of data manipulation stretching back to ancient times and it includes Newton. It has been claimed that "Up to half of the scientific papers in the United States may be contaminated by data manipulation." Universal Mendacity teaches to look beneath claims of honesty and credibility to find the opposite. Where academic advancement depends on publication, why should the paying public even suspect that published results, in whatever field, are not largely fabricated?

Perhaps there is a reason and this is the second objection. A feature of physical science that works in favor of honesty is the repeatability of physical experiments. If experiments are repeatable and if scientists are in competition, is not manipulation or misrepresentation open to exposure? Taking two contemporary claims as examples, high temperature superconductivity was widely accepted and favored by governments at about the time that another, altogether different, claim for cold fusion languished. In the first case, the original claim of Bednorz and Muller in 1987 was not only repeated but rapidly enhanced; whereas the interpretation of the second - with its hope for clean nuclear energy at laboratory temperatures - has been disputed alongside a failure to reproduce results. The first claim has been recognized by the noblest prizes; the second was published, unrefereed, and vigorously disputed. The first led to the greatest effort ever expended in developing a single type of material; the latter suffered funding strangulation.

The chief effect of repeatability of physical experiments is more profound than the resolution of disputes; repeatability aids the process of closure by which scientific assertions move towards acceptance and propositional status for further research. Nevertheless, the fact that experiments are repeated casts doubt on first results and if on one result, then on all results. Moreover, two measurements are never identical in all respects. Leaving aside misconduct, some differences are merely statistical and therefore insignificant. Other differences derive from peculiarities of experimental method and can sometimes be satisfactorily explained.

Measurements can always be made more accurately. Many countries support National laboratories that do just this. Whatever the measurements, experimental results and scientific laws remain approximations to truth and open to enhancement and question in the final analysis. Richard Feynman, the Nobel prize winning physicist, sounded this drum:

> "Each piece, or part, of the whole of nature is always merely an *approximation* to the complete truth, or the complete truth as far as we know it. In fact, everything we know is only some kind of approximation, because *we know that we do not know all the laws* as yet. Therefore things must be learned only to be unlearned again or, more likely, to be corrected." The italics are his or his transcribers'.

So any apparent closure in scientific law is opposed by the action of doing science.

The fact of an assertion is a part of its truth, and, in this truth, reality is opposed to the ideal intention. That is why, in the best laboratories, theories and assertions are treated skeptically by routine. Universal Mendacity is implicit in the behavior researchers find there. The theories and experimental reports are fired in the crucible.

Think of the criticism that Crick and Watson experienced in Cambridge before arriving at their double helix structure for DNA, the discovery that founded modern biological science, the discovery that explains the replication of species and that encodes evolutionary changes. Universal Mendacity seems to be affirmed by the series of false starts that progressed to the target discovery, but the bullseye may seem equally to be the exception that disproves the rule. What Mendacity can there be in that supreme 20th Century discovery?

The reality, at the time, was hidden. As we now know, a large body of new science, including genetics, sprung subsequently from the discovery. The discovery was a great step forward for three men and a woman as well as for mankind; but it was a step into the unknown, full of meaning, not all of which were opposed to the original intentions of the authors. When the assertion moved to closure, the discovery became propositional.

After a discovery, it is easier to know what research should have been done. Meanwhile, beyond issues of credibility, administrative bodies that are responsible to the public for its expenditure, will look for benchmarks and tests. In practice, the three executive rules are observed

47

by their neglect. How can they be applied to promote safer, more efficient and more productive science? The question recurs in the following chapter.

CHAPTER 4

WHO GUARDS THE GUARDIANS?

> There are ways of knowing, beneath what is said. A rise in voice pitch is suspicious and so is sweating. "Mr. Smith was looking sideways because he was telling a lie."

Success in cheating is the greater advantage in zero sum games than where cooperative behavior gives added value. Think of cards; think of marriage. A research director in a semiconductor wafer foundry taught the politically correct doctrine that the best technology wins. We start to lift the carpet at a date around 1940. This was the year in which the transistor was discovered, marking the beginning of the silicon age and information revolution. We will proceed to the future of chip making, specifically to the turmoil in selection of next generation lithography which will be critical to advances in the near future. You'll see how Universal Mendacity guards against competing claims.

Engineering is the lazy side of science: it is the way of making things with economical effort. In doing so, a recognition of Universal Mendacity is not a recommendation for dishonesty. Fun people seek maximum enjoyment from life, and that comes with understanding it. Sometimes you learn more from dishonest assertions than from truthful ones, or even bland ones. This is a story which ranges from science to manufacturing and in which there has been substantial waste of investment.

Our observation is more for the big spenders in scientific research. This is a natural concern for a public that funds large expenditures; but there is also, by contrast, a private side to science with its private or premature successes. Whether public or private, novel prediction and its verification form the criteria for successful science, as is widely recognized. One difference separating public science from its industrial application is the reference given by profit. This is a fact that regulates, even in the presence of ignorance or unfairness. The fact creates stepping-stones for competitive newcomers. A particular technology, that is the best engineering solution for getting things done, may not be implemented because of established interests. These know that if they do not protect themselves, they will not maintain leadership. Protection comes in miriad ways that include, at two poles, superior technology and restrictive practice by monopolies or cartels. Between these poles lies control through conglomeration with startups before they become rivals. In the end, the consumer pays.

Research laboratories, around the world, are limited by the support they get. Government grants depend typically on industrial co-funding. A simple change in company management can destroy a major laboratory. Research institutes and academic researchers conduct a game of survival. It is a game of deceivers and dupes. The semiconductor industry, which has led modern communications and which is therefore a cornerstone of modern culture, has many examples. Here is one:

For decades, in the development of silicon circuits, the number of transistors on a chip has doubled every two years. To this fact is due the large increase in the speed of microprocessor chips and in the storage of memory chips. The development has followed a shrinking of pattern dimensions. The shrinking occurred in the early stages by reducing the pattern sizes written onto masks, used like photographic negatives, in proximity printing. Later, the shrinking proceeded by a reduction in printed pattern size through demagnifying in projection. This is like printing photographs from a 35 millimeter negative film except that the print is smaller in lithography; not larger. Subsequently, to reduce diffraction smearing, short wavelength lasers were used: ultraviolet and deep ultraviolet. How to get further shrinks in the next generation? There are several ways, but maybe none will enter the factory and benefit the computer store. Chips could conceivably settle into a trough of limited dimension - like the jumbo jet of forty years ago - so that the power of silicon chips would plateau. Wouldn't that leave an opening for rival newcomers? Yes it would, but the manufacturing process is so complicated that entry is not easy.

What in fact happens in the melting pot is that a dialogue of wishes and fantasies replaces informed scientific discussion of relative merits. However disheartening for the scientists involved, it is the play of economics and company politics that decides the actual spending on development. Over many years, great effort has been spent on short-wavelength, X-ray systems that have proved successful at printing the finer dimensions. The X-ray methods make use of Einstein's theory of relativity to condense light that has produced batches of commercially valuable silicon chips. But the X-ray systems are relatively bulky only because, like electicity generating stations, each

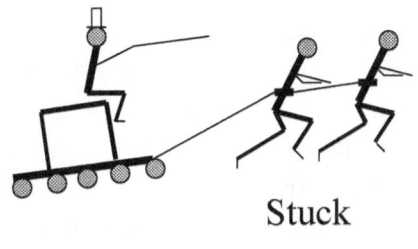

Stuck

Professional

Amateur

unit produces a large output. They are called non-granular. Manufacturers, unwilling to face necessary change, typically adopt a fantasy. They imagine an abolition of physical barriers in wishful modular systems: systems that might be swapped either when faulty or when future upgrades might be needed. To make way for the fantasy, the X-ray development was curtailed. For a cost competitive industry, development of a star wars pseudo-solution was funded instead, one that coincided with a need for defense laboratories to find civilian work. Unfortunately the star wars solution is extremely short of light, does not take advantage of relativity, is narrow band and is excessively costly. Shoot turkeys with an F18 and they'll die, but you won't buy them in Safeways. Will you buy a star wars PC five years from now? The best technology does not always win.

The fantasies are aided further in many ways by researchers and business units that might know better. In the turmoil, the X-ray methods were naturally misrepresented: they were wrongly projected as "one-generational" when, in fact, they had the widest range of all contenders, besides being the only method properly demonstrated with proper cost control. Truth was sacrificed for policy. Technology development in a vital industry was curtailed for scientifically

Thanksgiving

Professional

Amateur

unsupportable industrial politics. Universal Mendacity teaches that, since when claims are made the truth is never simple, there is an opposite meaning that weighs heavily. There is, moreover, a law of proportions. The greater the error, the more ridiculous; but the predicted squeeze in the cost of cutting-edge manufacture will see a bitter fruit in reduced output and the product of decisions that are technically wrong - wrong in engineering because of excessive complication and wrong in physics because of insufficient light.

The semiconductor industry is extraordinary: one well known company spent a billion in developing a perfectly good technology and then abandoned it. Why? To make it easier for another well known company to commit the same developing a perfectly bad technology? Is it not surer that the bad technology, called EUVL, will fail than that the good technology will be revived? How come the initial investment was made, including public funds, if it was not needed? It is a sad fact that engineers without physics are actors without playwrights and furthermore that the burden of the technical errors fall most heavily not on the persons responsible for it; but on persons outside the company: on equipment suppliers and on smaller laboratories. Of course, investment decisions are complicated, and within companies there are diverse voices that compete for attention. Cornering markets is one thing; public funds another. A weaker challenge to the established, revenue producing, optical methods would come from next generation also-rans. The most fruitful way of understanding the claims is through Universal Mendacity: believe not what they say; but observe how they say it and know why they say it. We'll come to belief later.

What does context imply for stories like this?

a.   People - whether technical or not - make promises when they can't verify.

b.   Reality opposes claims, otherwise they would not be significant. The simpler the claim, the more it opposes.

c.   A process in hand is worth two in the International Technology Roadmap for Semiconductors.

There are further implications. Why does peer review within the corporations that fund development not identify true weakness and real opportunity?

Peer review is the principal method used, around the globe, in the administration of both scientific publication and of research funds. Take publication first, while noting that the implications apply even more significantly to the allocation of research funds. Historically, publication in a peer reviewed scientific journal has been necessary for the general acceptance of a theory or piece of evidence. Authors often express annoyance with reviewers, but a scientist who has revolutionary results does not act rationally if he relies on peer review for acceptance. Often it is the reviewers themselves who failed. They are not naturally selected for impartiality, and so review by established authors provides a bias opposing novelty and change. Typically the reviewers are held secret by journal editors so that no responsibility is attributed. Open mechanisms for rebuttal and justification are not available. It is difficult to expose abuses in the journal refereeing system owing to its secrecy, and the structure of the enterprise does not penalize endemic and discreet misconduct. Ironically, there are modern challenges to the systemic abuse not only due to its inherent weakness, but due also to the information explosion on the internet. The most rapid publication is now not refereed at all, so that scientists can happily bypass the old refereeing procedure. They use their own procedures for discriminating reliable information from unreliable. Not for the first time will the internet bring a democratic liberation from obstructive control.

In the restricted community of scientists, secret review does not advance creative and unbiased behavior. On the contrary, the incentives point towards conservative self-preservation, especially of anonymous reviewers. Following the assumption of executive rule 3 (unenlightened self interest) scientists are no more honest than corporate America. Then the simple rational executive solution is to make the reviewing process, at the same time, public, accountable and credible. Unfortunately, it is within the interest of established scientists to see that enlightenment does not occur, and open review is unlikely to happen without applied external force.

There is a force: without reviewing, newspaper reports are sometimes needlessly disruptive for the public. Examples range from topics like vaccination side-effects, new drugs and procedures to the wider fields of science and engineering. It is predictable that the longer reviewing remains secret, the more will unreliable reports reach the press and public.

Some disciplines outside mainstream science do reveal reviewers' identities. Within personal memory the Times Literary Supplement changed from secret to open reviewing of books. Complaints of bias are, even now, not uncommon, and indeed should be expected, but the process is made transparent and bias can be exposed and left to readers to evaluate. Of course the editor retains control of what is published in his bi-weekly journal, but if open reviewing finds objections from the reviewers of other journals, notice that there professional ambition ensures reviewers are not in short supply, and are therefore replaceable.

The reviewing process for research proposals is even more secret. This is due to the obvious fear authors have that their plans might be stolen. The problem is to find either external tests or independent auditors who are able to weigh the arguments for and against a proposed scientific endeavor. How can bias in testing be minimized?

Some results of science are spectacular: the light bulb, nuclear power or the computer. Research results become manifest, typically, after a long period of incremental research. But there is a difficulty in using claims for future results as benchmarks for the incremental growth that is typical of scientific research, "Normal Science" as Kuhn calls it. The results occur long after the investment decisions have been taken. How can tests be arranged so that further spectacles can be funded and invented before the inventions themselves are made?

The art of managing science may be well developed somewhere, but the principles given in executive rules 1, 2 and 3 are observed by omission. There are ways by which conflicts of interest in the constructive use of the resources might be overcome. The same process would protect the public from environmental aspects of scientific practice. For efficiency, a working structure is needed that takes due account of ignorance, career goals and self-interest in both administrators and practicing scientists. An administering body needs to employ sufficient knowledge combined with sufficient independence. It should include knowledgeable scientists and exclude researchers in any way dependent of the programs they review. For sustainable scientific development with big spending, the enterprise must ultimately produce profit. To see how the various goals might be achieved, we consider a scientifically hierarchical pyramid that is inverted, decisions being taken top-down, from the broad applications base to theoretical apex. We begin by supposing that all major disciplines are, as in the past, supportable in principle.

Using this pyramid, *qualifying tests at each level are found higher in the inverted hierarchy.* For pure mathematics, the qualifying tests

are

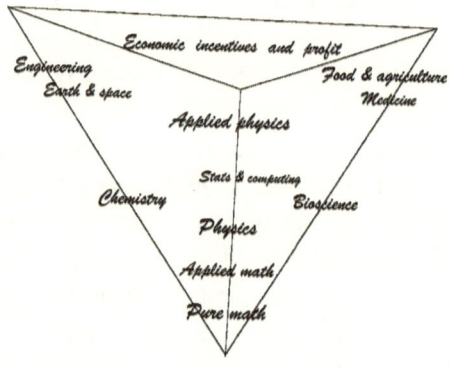

success in applied mathematics; in applied mathematics the qualifying tests are found in the physical or life sciences; for physics, they will be in applied physics and engineering; for engineering, they will be financial; for biological science, the regulatory tests will be medical or agricultural and so on. Success in application, or projected application, would provide independent management benchmarks for the quality of a particular scientific effort. Universal Mendacity could then be applied at one remove from the authors making claims. The inverted pyramid is democratic because control, including safety, will lie ultimately with the people affected. The pyramid would help protect the paying public at the same time as improve accountability in the use of its funds.

The trouble is, of course, that regulation of science by this type of hierarchy of tests has already been tried. In parts of the world that are deficient in scientific learning and where this is valued only for its pecuniary usefulness in the manufacture of trade goods, the inverted pyramid is in fact directed top-down. Unfortunately, those are the parts of the world where the worst abuses occur, including environmental abuses. When there occurs a mix of ignorance, self-interest and occasional corruption, a rational, productive and safe choice becomes improbable.

Nobody guards the guardians and, on a broader front, that is why truthfulness, in an age of weapons of mass destruction, is a matter of public survival. Ranging to the microcosmic, that is why, realism and the recognition of Universal

Mendacity are necessary for individual survival: especially for the poor young men, on whatever side, who do what they're told and risk their lives without sufficient question. On its own, the inverted pyramid does not, as examples show, assure the public of credible and safe management. More is needed. Voters and administrators would both be misled from ignoring Mendacity.

CHAPTER 5

MENDACITY IN BELIEF SYSTEMS

To one side of assertions lie propositions; to the other beliefs. Religion owns the additional complication of unscientific myths that shape the intention and mold a part of reality. Its music is part divine like its hymnology and poetry - features displayed, to varying degrees, in all religions.

A long avenue of freshly green trees
Leads up to great, gray, gaud, unfolding clouds,
The raindrops on the gently swaying leaves
reflecting, mirror-like, a now sparkling sun.

And I asked how it came to be
And why it should be so
Above all, why we
are so blessed to perceive?
For no real reason philosophies abound.
Perhaps it is just, just, just so.
The world goes round, goes round, goes round.

Can that be serious, "for no real reason?"
Why think and write and sigh?
Why laugh and talk and cry?
Why play and walk and fly?
Above all why the why?

If you consider this last question dumb, you can skip this chapter, which is not written for you. Some questions have no answer. Only notice that the answer, in whatever form, consumes the lives of many others, especially those who love and savor life, marveling at their own existence. Maybe Neitzschean aphorism is best:

If, from whatever reason or motivation, I say I believe in God, I imply something about the opposite and vice versa. Fortunately, whatever I say, and if you can read this, God must believe in me - that is if He is the ground of our being and the foundation of our contingency. (I hope not to be penalized for this assertion on the last day as it seems reasonable enough - doesn't it? Traditionally, God is Absolute Reason.) Bertrand Russell argued against our experience of contingency. At the pearly gates, he said, he would ask, "Why didn't You make Yourself more evident?" Only if we say nothing, is nothing said.

If you see an elephant you can shoot. If you don't see it, you don't shoot (whether with a click or a rebound). If god were dead, why would you say so? To bring Him to Life?

In history there are many gods and many meanings for god. Beyond the god of the philosophers are the gods of the various religions, many thought to uniquely reveal the One true God. Some religions are dogmatic, teaching a body of simplified doctrine; other religions are open and inclusive; many are both. Under Universal Mendacity, there is nothing so much like an atheist as someone who professes belief in everything. For every dogma, there is an opposite: church history shows this in the debates that precedes new dogma. Dogmas are like physical laws: they map a belief system and are stepping- stones for progressive development. The reality is unfathomable.

When a church says that the existence of God can be proved by reason, you ask, "How, where and when?" Why is the proof not universally accepted - Like propositional Newtonian gravity? If the proof were shown to be valid, why would it be necessary to claim validity? If not, what is implied of the belief system? That it believes what it knows to be dubious?

If religion were about what is beyond rational argument, then arguing about it would be like arguing over taste. Actually, of course, the subject is different. There are as many ideas of what religion truly is, as there are ideas of God. Some parts may be rational and some not. But if belief depends on experience, then what difference does its metaphysic have for behavior or happiness?

If, on the other hand, it could be proved that belief is irrational, you could choose belief and remain rational. Truly rational systems are deductive, like abstract geometry, and consistent. By contrast, Universal Mendacity provides space for opposite assertions of belief: expound one doctrine and you know that there is a real opposing meaning that made you do so.

All religions provide moral teaching and this can be more rational or less. That is, good actions can be analyzed consistently with accepted rules, or if less rational, asserted with inconsistent reasons or even with no reason. Some teaching is uncompromising on honesty as in this statement by Aquinas:

> "All lies are by definition wrong, acts out of tune with their matter, since words should by definition signify what we think, and it is perverse and wrong for them to signify what is not in out mind."

The great religions have always symbiotically nurtured philosophers: the medievals, Averroes and Avicenna, Maimonides or Aquinas, came from various belief systems, all of which have continuing influence in the modern world. There are, in religion, acres for profound and unspeakable belief as there are for uncompromising simplicity. Surprising, is it not, that we use one word to describe both of these and all that lies between?

Bo De Yang

CHAPTER 6

CORPORATE AMERICA

You can skip this too if you like. It is where honesty and deceit makes and breaks.

For every successful act of crookery there is a dupe. Universal Mendacity teaches how to avoid being duped in a way that does not suffer from unnecessary loss of communication. You are not required to be more credulous, skeptical or cynical than circumstances demand or permit. You need always to study the reality that underlies assertions. You need to know about the opposite meaning that subsists in the significance of assertions and to distinguish their opposed meanings from the truth or falsity of simple propositions.

Where an honest appraisal of reality is important, complications arise from duplicity. Why should anyone call you too honest if he is not engaged in deceiving either himself or you? If the reality is that *he* is not honest enough, you'd like to know what he is hiding and why he is hiding it. If he should, by abstraction, be stating a proposition, remember that these are either right or wrong. Nobody can then be too honest. But with regard

71

to the assertion itself, the real context and ideal intention are opposed.

Some cultures are more honest than others. In Victorian England, a gentleman's word was his bond, but he was supposed not to question the superiority of his class. He depended on a social system that he individually reinforced. You could trust his word, but you couldn't trust his politics. His apparent honesty was a tool for implementing his political superiority. Rulers require honesty of subjects when they need information, but if they implemented the requirement themselves, they would sacrifice their power to the consequences of too honest appraisals. That is why lying to the house was worse for a British Secretary of State for War than lying in the nude. Forty years later, there were resignations after Iraq was invaded in 2003. On the other side of the Atlantic, in the Wild West, honesty is held in scant regard. If lying on oath was impeachable by a senate majority, the charge was rather an ineffective exercise of power than one of deviant official behavior. The peccadillo pales besides the deceits of prior presidents especially in matters relating to war: Polk on the Mexican attack, Theodore Roosevelt on Colombia and many others subsequently. For these, deceit was a means for exercising their wills, clearly a prerogative of executive authority wherever there are dupes for the hearing. A Western public tolerates '45 minutes' in presidential addresses.

Some deception in appropriate circumstances is normally regarded as professional. Doctors who report to their patients unvarnished diagnoses are called insensitive. Physicians employ sophisticated rituals for terminal illness. An intelligent patient can see through doctorspeak and prepare with style for the worst. For a different profession, Williams gives the example of equivocation by the Jesuit Garnet, referred to in Shakespeare's Macbeth, and who died believing himself a martyr. A priest may not reveal what he has heard in confession and may not reveal that he has heard the confession. He "does not know anything" that his questioner may rightfully learn, so that if he is asked if he knows anything he should say, "No." Others, with the intention of honesty, would try to evade the unwelcome question, but the evasion is itself a non-verbal assertion. In Universal Mendacity any answer will give some information. The best answer is the one with the most desirable consequences, as understood.

Horse traders, car dealers and stock analysts know this implicitly. In the buyer's view, a good dealer will sell not the car they want to be rid of, but the car you need. He survives economically because he sells repeatedly to loyal customers. By contrast, when a stock analyst writes "buy" to enhance his commission, his reader would do better to sell. It is not necessary to demand absolute honesty in order to avoid being a dupe. The avoidance is within individual control. There are many choices, and recent history raises particular problems for company shares - especially for the small investor. Why should anyone give money to a board of directors, to do with more or less as they please - unless for such incentives as a modern board of directors would not part with? The individual investor has to take the system entirely on trust. When he invests in publicly quoted shares in large companies, he loses control of his savings to boards of directors and mutual funds, to do battle with on terms of their own.

Doubtless, there'll be tinkering with the supervisory structures that oversee investments and company practice, and doubtless the tinkering will be copied in many countries; but the fundamental problem will recur if it is one of credibility and of dupes. That is because credibility is not easily restored after it has been lost. But why should the problem occur in one sector, corporate practice, without imbuing all public life in civil America: in state and federal government, in charities and education, in professions and in trade? Where honesty is absent so will communication fail: in collaborative research, commercial success, the progress of technology and the progress of culture. Universal Mendacity is a way to recover the crumbs and to compress a meal of them. Here's a tip: without it, you'll lose.

Bo De Yang

CHAPTER 7

CONCLUSION

If you objected to this book, you verified it: you noticed the opposite. Nothing here stated lies outside the scope of Universal Mendacity. In the idea of the assertion, the Real and the Ideal are united. There is always more to say, so everything we write is therefore naturally tentative.

If you didn't object, were you conned? In between is an unwritten reality that we are ever intent on describing with words.

Others will say there is nothing new here. Like Plato's midwife, they knew all along what every back-street boy knows: you don't believe what you're told. Unfortunately, books are not written in back-streets. Writers often think words have an independent value like the monetary gold standard. It's not used now.

Our world may not last the Century that has started so violently, but survival of civilization will be less likely for not traveling the ridge atop the destructive precipices of demagogic credulity and absolute cynicism. What to believe of

77

ambitious people who acquire power over their fellows by tricky means? That they say what they must to get their way? Universal Mendacity teaches not to believe what they say, but to notice the way they say it and to know why they say what they do.

REFERENCES AND NOTES

Page vii. Dr. David Kelly: was a scientist who investigated Iraqi weapons. He was found dead after he had been named for giving information to the BBC and grilled by a parliamentary committee. Soon after, the prime minister's media chief resigned.

Page viii: NASA: The National Aeronautical and Space Administration was responsible for the loss of two space shuttles that ended in disaster with the loss of their crews.

Page viii: *Our Final Hour:* by Martin Reese, Princeton 2003.

Page 1: Bernard Williams: *Truth and Truthfulness,* Princeton 2002.

Page 9: Propositions: Propositional knowledge has a long history extending from Plato Theaetetus 201c-210d to G. Ryle 'Letters and syllables in Plato', *Philosophical Review,* 1960 pp 431-51, for example. If, for Plato, propositions are 'true knowledge'; for us they are an abstraction from reality.

Page 10: Peace for our time: the assertion made by the British Prime Minister, Neville Chamberlain, after the Munich conference and prior to the German invasion of northern Czechoslovakia in 1938 that preceded the Second World War.

Page 10: The *real* meaning came from the context: for the importance of context see J.L. Langshaw, in *How to do things with words* 2nd ed. Oxford: Clarendon Press, 1975.

Page 12: "Lord I believe, help my unbelief." Mark ch 9 v 24.

Page 14: intuitions developed at critical moments: as in R.M. Hare, *Moral Thinking, its Levels, Method and Point,* Oxford, 1981.

Page 14: It will learn to discriminate innate signs of bad intentions: see A Vrij *Detecting lies and deceit: the psychology of lying and implications for professional practice* 2000 Wiley.

Page 16: He LIES and he knows he LIES: from James Boswell *The life of Samuel Johnson* Book V.

Page 16: There is a sociological literature on lying: for a bibliography, see J.A. Barnes, *A Pack of Lies*, Cambridge University Press 1994.

Page 19: the *esse* of material things is *percipi*: to be is to be perceived. see G. Berkeley, *A Treatise concerning the Principles of Human Knowledge.* Subsequently the philosopher I. Kant, in his *Critique of Pure Reason,* described a duality between ideal thought and real phenomena where the causal 'noumena', things in themselves, lie beyond knowledge.

Page 19: In the idea of the will the real and ideal are united: G.W. Hegel, *The Science of Logic.*

Page 21: If it is a member of itself it is not; and if it is not, it is: Russell's paradox as described in his *Autobiography*, Routledge 1998 and in A.N. Whitehead and B. Russell, *Principia Mathematica*, Cambridge University Press, 2nd ed, 1927.

Page 21:　　K. Godel: *Monatshefte fur Mathematik und physik* Vol 38 1931 pp 173-198. Or see *Godel's Proof*, by Nagel and Newman, Routledge and Kegan Paul, 1971.

Page 22:　　Cynics: Antisthenes and Diogenes of classical times may not be examples; we refer to a more common error of too easily finding fault in others.

Page 22:　　I cannot tell a lie, I didn't do it: Richard Nixon, on the Watergate. When Washington used similar language, he followed by *admitting* to having chopped his father's cherry tree.

Page 24:　　logical positivism: something like A.J. Ayer, *Language, Truth and Logic,* 1936.　　He subsequently developed the themes as did others.

Page 24:　　the scientific method operates by disproving theories: K.R. Popper, *The Logic of Scientific Discovery,* Hutchinson 1959.

Page 26:　　*The Structure of Scientific Revolutions:* T. S. Kuhn, Chicago University Press, 3rd ed. 1996.

Page 29:　　mathematical tautologies… following experimental evidence: W.V.O. Quine, in *Methods of Logic* 1950, *From a Logical Point of View* 1953 and Word and Object 1960, argued against a sharp distinction between analytic truths and synthetic truths but we do not need to examine that here.

Page 29:　　discoveries of these laws: T. S. Kuhn, *The Structure of Scientific Revolutions,*, Chicago University Press, 3rd ed. 1996.

Page 30:   J. Magueijo" *Faster than the Speed of Light,* Perseus 2003. Books, that get published with such strong criticism of established reviewers, are not common.

Page 31:   the Michelson Morley experiment: details can be found in many physics text books.

Page 32:   Ernst Mach: physicist and philosopher, realized that space and time do not map onto a Newtonian grid, but depend on gravity in the existing universe.   Charles W. Misner, K.S.Thorne and J.A.Wheeler, *Gravitation,* Freeman 1970, p. 411.

Page 33:   The biggest blunder of my life: for Einstein's remark see Misner et al. in the last reference (for page 32).

Page 33:   serve usefully in current speculation: see Martin Rees, *Our Cosmic Habitat*, Princeton, 2001.

Page 34:   probability amplitude: strictly the intensity, or the product of the probability amplitude with its complex conjugate.

Page 34:   foundations of quantum mechanics: a more detailed discussion is given by M. Rees in *"Our Cosmic Habitat"Princeton University Press, 2002.* Gravitation and quantum mechanics have still not been unified. See also A Pais, *Subtle is the Lord, the science and life of Albert Einstein,* Oxford University Press 1982. For a general account see *Quantum theory and the schism in physics,* KR Popper, ed WW Bartley III Hutchinson 1956. An example of a detailed objection can be found in A Einstein, B Podolsky and N Rosen, *Can Quantum-Mcahnical Description of Reality be considered Complete?* Physical Review **47** (1935) p 777.

Page 36:   exploring new territory: there is in fact more to experimentation than is described in K.R. Popper's *The Logic of Scientific Discovery,* Hutchinson 1959.   Some scientists disprove theories (think of parity violation by Wu); more of them chart new territory or take advantage of serendipity (think of the microwave background in space).

Page 38:   health hazard: an example is the Singapore Synchrotron Light Source, though it did produce intellectual property before it became operational.

Page 39:   *Those who know, don't fight*: at least they don't fight for long if they are like Dr. Kelly, reference. page vii.

Page 40:   Northcote Parkinson Cyril, *Parkinson, the Law*, Houghton, 1979. the inventor of Laws that include "Work expands to fill the time alloted," and "Administrators rise to their level of incompetence."

Page 41:   *the Concept of Law:* H.L.A. Hart 2nd ed. Clarendon 1997.

Page 41:   The Institute of Materials Research and Engineering, Singapore. One example: in some places, science research is impossible. Individual institutes may be capable of reform.

Page 43:   biased towards truth and truthfulness: For an claim of bias for honesty in academia see B. Williams, reference for page 1.

Page 44:   includes Newton: see J.A. Barnes, *A Pack of Lies*, Cambridge University Press 1994.

Page 44:   contaminated by data manipulation: O'Neill, Graeme 1991 Truth hurts US scientific 'facts'. *Age (Melbourne) 5 June [56]*.

Page 45:   high temperature superconductivity: Tinkham Michael, *Introduction to Superconductivity*, McGraw Hill 2nd ed. 1996 or A. Bourdillon and N.X. Tan-Bourdillon, *High Temperature Superconductivity - Processing and Science,* Academic Press, 1994. It is remarkable that these books have not been overtaken by subsequent developments.

Page 45:   favored by governments: the US Internal Revenue Service makes special allowances for the development of high temperature superconductivity.

Page 45:   J.G. Bednorz and K.A. Muller: *Zeitschrift fur Fusik B* **64** 189, (1987).

Page 46: Richard Feynman: worthy support for this view is found in R.P. Feynman, R.B. Leighton and M. Sands, *The Feynman Lectures on Physics,* Addison-Wesley 1964, vol I section 1-1.

Page 47:   Crick and Watson: J.D. Watson, *Double Helix - a personal account of the discovery of the structure of DNA,* Atheneum 1968.

Page 49:   Mr. Smith was looking sideways because he was telling a lie: children can read about the Great Blackhand Gang in Grahame Greene's *The Little Old Steamroller.* A priest's homily notes were found inside one copy. In the margin was a squiggle: "argument weak - shout." Further down, a word was crossed out and replaced. The uncorrected phrase was, "Cheaters better prosper," which wouldn't have done in a homily, however true - as true in the industry of science as elsewhere.

Page 49:   Cooperative behavior: R.Axelrod, *The Evolution of Cooperative Behavior,* Basic Books, 1984.

Page 52:   X-ray systems: Thomas Jefferson National Accelerator Facility, VA, Technical Note 03-016 (2003).

Page 54:   star wars solution: Extreme UltraViolet Lithography, however much favored for by business, suffers extreme physical disadvantages. It does not have enough light intensity: it is narrow band; it does not use a relativistic condenser; and its optics are extraordinarily complex, requiring aspherical multilayered mirrors machined and coated to a minute tolerance and subject to contamination. With military style lasers, this is the star-wars solution described above.

Page 56:   EUVL: extreme ultraviolet lithography, actually projection X-ray.

Page 57:   Next generation also-rans: as exercises in obstinacy, it will be accidental if Immersion and Imprint lithographies do not disappear in the same slow cloud as SCALPEL (scattering limited projection electron lithography) and EUVL (extreme ultraviolet lithography) - each for its own particular reasons.

Page 58:   Authors often express annoyance with reviewers: for example J. Magueijo, *Faster than the Speed of Light,* Perseus 2003.

Page 60:   "Normal Science": T. S. Kuhn, *The Structure of Scientific Revolutions,* Chicago University Press, 3rd ed. 1996.

Page 66:   experience of contingency: Bertrand Russell, *The Existence of God,* ed. R.Hicks Macmillan 1964.   The following "pearly gates" remark was heard on a BBC interview.

Page 68:   Aquinas, *Summa Theologiae* IIa IIae, Q69, 1 and 2, English translation Gilbey (1975) 117-119.

Page 72:   a gentleman's word: The epithet of gentleman is dubious; one that belonged wrote: "A gentleman is to be defined as one of a society of equals who live on slave labour, or at any rate upon the labour of men whose inferiority is unquestioned."   Bertrand. Russell, *History of Western Philosophy.*

Page 72:   Dr. John Profumo was the Secretary of State for War who resigned from Harold Macmillan's administration in 1960's England.

Page 72:   resignations: the first was Mr. Campbell, a senior advisor to the British prime minister, resigned while an inquiry was being conducted by Judge Hutton into the death of the scientist, Dr. Kelly.

Page 72:   ineffective exercise of power: In 1999, the US congress voted to impeach president Clinton; but, on party lines, the corresponding vote in the senate fell short of the two thirds majority that would have been needed.

Page 73:   '45 minutes': the time reportedly needed for Iraq to unleash weapons of mass destruction abroad.   It was read in the presidential address, was a cause for war, and lacks public evidence, neither found in existent weapons nor in reliable secret information.

The Law of Universal Mendacity
- and don't be conned

Bo De Yang

About the author

The author is an established writer, practicing scientist, professor, laboratory director, and company executive. He was trained first in philosophy, and he brings a new and thoughtful analysis to the hardheaded world of survival. He applies particular knowledge of the materials used in the high technology semiconductor industry and to the structures needed to provide safety in an ever more nuclear world. You don't have to believe his claims because you've got the book.